Earth: Measuring Its Changes

Table of Contents

by Barbara M. Linde

Introduction 2

Chapter 1 Earthquakes 4

Chapter 2 Glaciers 13

Chapter 3
Water and Wind Erosion 22

Conclusion 29

Glossary 31

Index 32

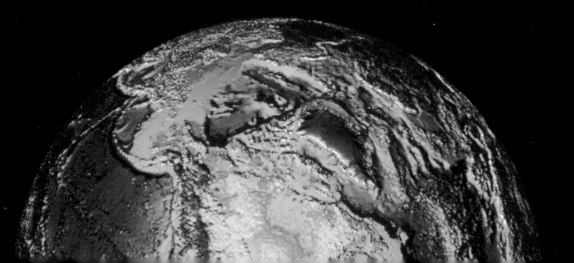

Introduction

Our Earth is changing all the time. What are some of the **forces** that cause these changes?

Earthquakes (ERTH-kwakes) can change Earth's shape. They can cause a lot of change very quickly. **Glaciers** (GLAY-sherz)—huge masses of ice and packed snow—move rocks, dirt, and sand from one place to another.

Wind and water can cause **erosion** (ih-ROH-zhun). Pounding waves move sand and change shorelines. Wind moves sand and dry earth away.

◄ The Lambert Glacier in Antarctica is over 200 miles (320 kilometers) long. It is one of the largest glaciers in the world.

▲ sand dune in Namibia

Earthquakes, glaciers, wind, and water are all forces that change Earth. Scientists can measure these forces. They measure the strength of earthquakes. They measure how fast glaciers move. They measure how quickly erosion takes place.

How can scientists measure these forces? What are the forces like? Read this book to find out.

CHAPTER 1

Earthquakes

To understand earthquakes, you have to understand the structure of Earth. Earth has four main layers. The inner and outer cores are deep inside Earth. The mantle (MAN-tul) is around the core. The mantle is mostly solid rock. But its top part is hot, melted rock, about 1,800 miles (2,897 kilometers) thick. Earth's crust floats on top of the mantle.

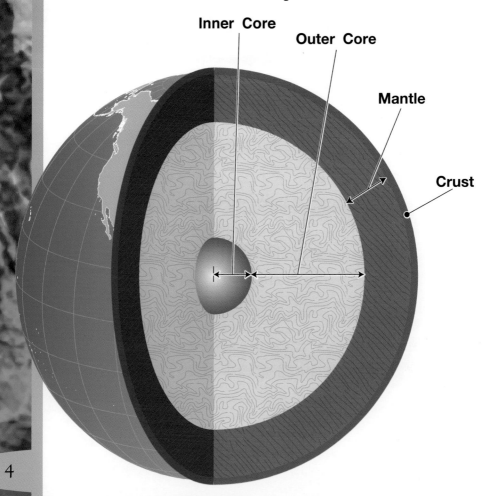

Earth's crust is broken up into giant **plates**. The plates are like huge, flat boats floating on Earth's mantle.

▲ Earthquakes take place at the edges of Earth's plates.

The moving plates send out energy. Sometimes the plates get locked together, and the energy cannot escape. The force of the trapped energy becomes so strong that it pushes the plates apart. The energy travels to Earth's crust. It shakes everything in its path. This shaking movement is an earthquake.

Sometimes, Earth's crust will crack where an earthquake has happened. The crack is called a **fault**.

It's a Fact

There are more than 1,000 earthquakes around the world every day. Most of them are too small to feel. Only about 150 earthquakes per year are strong enough to feel.

▲ Many earthquakes occur along the San Andreas Fault in California.

CHAPTER 1

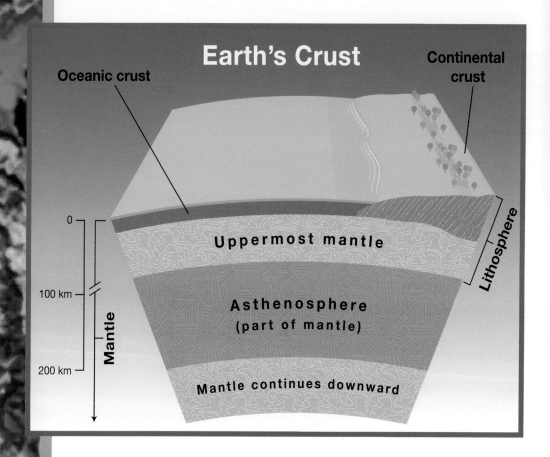

The point on Earth's surface that is the source of an earthquake, or where the earthquake starts, is the **epicenter** (EH-pih-sen-ter). The area around the epicenter has the greatest energy and the strongest shaking.

How Are Earthquakes Measured?

Scientists called **seismologists** (size-MAH-luh-jists) measure the energy of earthquakes. They use a **seismograph** (SIZE-muh-graf). It picks up and records the vibrations that radiate out from the source of an earthquake. The recording is called a seismogram.

Scientists place seismographs in or near faults. They take measurements day and night. Each machine is connected to a large computer system. Scientists use information from seismographs to locate and measure earthquakes.

▼ These machines are seismographs. The word part *seismo* means "about earthquakes." The word part *graph* means "write."

CHAPTER 1

1. Solve This

Here are three seismograms (SIZE-muh-gramz) from an earthquake that happened in Alaska on October 23, 2002. The earthquake was recorded on seismographs around the world.

How much longer did it take for the shock waves to reach Australia than to reach Massachusetts?

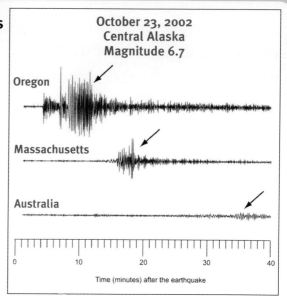

Math ✓ Point

What information from the seismograms do you need to solve the problem?

EARTHQUAKES

To compare the strength of earthquakes, scientists use a special scale, or measuring tool. It is called the **Richter** (RIK-ter) scale.

The Richter scale has numbers from 0 to 10. The numbers stand for the **magnitude** (MAG-nih-tood), or strength, of earthquakes. The least powerful earthquake measures 0 on the Richter scale. Each number on the scale represents an earthquake that is ten times stronger than the number just below it. So an earthquake that measures 1 is ten times stronger than a magnitude 0 earthquake.

> ## 2. Solve This
>
> A magnitude 5 earthquake is how many times stronger than a magnitude 3 earthquake?
>
> ### Math ✓ Point
>
> What information from the text do you need to answer the question?

Earthquakes smaller than magnitude 3 are mostly not felt. Earthquakes above magnitude 5 change Earth's crust.

They Made a Difference

Dr. Charles Richter developed the Richter scale in 1935. The Richter scale has been used to measure earthquakes on the moon and Mars!

a seismograph reading ▶

CHAPTER 1

Scientists also study the very slow movements of Earth's plates. This helps them know where and when an earthquake might occur.

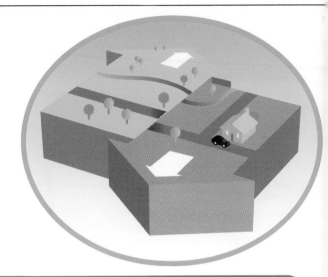

Careers in Science

Geology is the study of Earth, its rocks and landforms, and its changes. Some geologists study earthquakes. They might walk through the mountains for three or four weeks. They use seismographs to find places where earthquakes might happen.

▲ Geologist Dr. Avouac reads recordings from a seismograph.

EARTHQUAKES

▲ Laser beams can measure tiny movements along a fault.

CHAPTER 1

A Terrible Earthquake

On December 26, 2004, a magnitude 9.3 earthquake shook South Asia. Its epicenter was on the sea floor, deep in the Indian (IN-dee-un) Ocean. The earthquake caused a **tsunami** (soo-NAH-mee), a giant ocean wave. It moved up from the ocean floor.

Huge, strong waves came ashore in many countries along the Indian Ocean and the South China Sea. The waves also spread thousands of miles to the coast of Africa. The waves covered small islands and moved tons of sand and soil. The waves changed the shape of the land—all very quickly.

Read More About It
Read more about the tsunami at your school library or local library. Find out how people around the world have helped tsunami victims.

The top photo shows land before a tsunami. The bottom photo shows the land after a tsunami.

CHAPTER 2

Glaciers

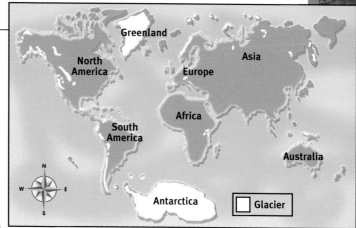

Glaciers cover almost 10 percent of Earth's land. Most of Earth's glaciers are in Greenland and Antarctica.

Some of these masses of ice and snow are about the size of a football field. Others are as much as 60 miles (100 kilometers) long.

Huge glaciers called ice sheets cover the whole continent of Antarctica (ant-ARK-tih-kuh) and most of the country of Greenland.

It's a Fact

The ice sheet on Antarctica is more than 2 miles (3 kilometers) thick.

CHAPTER 2

3. Solve This

One meter is equal to about 3 feet. Suppose a glacier is 60 meters thick. About how thick is the glacier in feet?

Math ✔ Point

Is your answer an estimate of how thick the glacier is? Or does your answer tell exactly how thick the glacier is? How do you know?

How Do Glaciers Form?

Glaciers form in high mountains and at the North and South Poles. These places get a lot of snow in the winter. They are cool in the summer, so some snow does not melt. Each winter, new snow falls on top of the old snow. The weight of the new snow packs the old snow down.

In time, the layers of old snow change to solid ice. The ice becomes thicker and thicker. These changes go on for hundreds or thousands of years!

▲ A glacier becomes heavier and heavier as the ice builds up in layers. The great weight causes a glacier to move.

14

GLACIERS

▲ Glaciers on Alaska's College Fjord

How Do Glaciers Move?

When a mountain glacier gets very thick and very heavy, it starts to move forward. Sometimes there is water under a glacier. The water can cause the glacier to slide down the mountain. If there is soft dirt under the glacier, the glacier can slowly slide down on the dirt.

4. Solve This

Glaciers move at different rates. Suppose one glacier moved down a mountain at 98 feet (30 meters) per year.
a. How much would the glacier move in two years?
b. How much would the glacier move in 6 months?

Math ✓ Point

What strategies did you use to answer these questions?

CHAPTER 2

How Are Glaciers Measured?

Scientists measure how glaciers grow and melt. If they melt too quickly, they can cause floods and other problems.

Here are some other ways scientists measure glaciers:

1.

▲ Scientists can use a laser aboard a small airplane to measure the height of glaciers.

GLACIERS

2.

◀ A scientist uses lasers to measure the height of glaciers in the Canadian Rockies. In two years, he will measure these glaciers again. This way, he can tell how much they have changed.

3.

▲ From space, special cameras aboard satellites (SA-teh-lites) take pictures of and measure glaciers. Scientists compare the measurements and photos. They can tell if and how glaciers are changing.

5. Solve This

Suppose a glacier melted 12 inches (30 centimeters) in one year. If it melted at about the same rate each year, how much would the glacier melt in 50 years? Give your answer in inches and in centimeters.

Math ✓ Point

Is your answer the exact amount the glacier would melt in 50 years, or is it an estimate? How do you know?

17

CHAPTER 2

Scientists also measure how far and how fast glaciers move. In one method, they put a long stick in a rock near a glacier. Then they put rows of sticks into the glacier ice. These sticks move as the glacier moves.

After a few months or even a year, the scientists look at where the sticks in the glacier ice are now. They can tell how far and how fast the glacier has moved. The scientists repeat the measurements every few months for a year. They get a history of how the glacier moves.

6. Solve This

Glaciers that move quickly are called galloping (GA-luh-ping) glaciers. The fastest galloping glacier moved 26,896 feet (8,200 meters) in 82 days. The glacier moved at the same rate every day.
a. How many feet did the glacier move in one day?
b. How many meters did the glacier move in one day?

◀ The fastest galloping glacier is in the Himalaya (hih-muh-LAY-uh) Mountains.

Math ✓ Point

How could you check your answers?

▲ The Matterhorn (MA-ter-horn) is a mountain in Switzerland. Glaciers carved its strange shape.

Glaciers Cause Erosion

Glaciers are hard and very heavy. Glaciers can scrape the mountains. They loosen bits of earth and rock called **debris** (deh-BREE). The debris moves along with the glacier and helps the glacier cut into mountains and make deep scratches in rocks. This is one kind of erosion. It can change the shape of mountains.

Sometimes heavy glaciers dig deep holes into the soft rock under them. When the glaciers melt, the holes fill with water and become lakes.

▲ It took thousands of years for glaciers to carve the bed of this lake.

Glaciers covering Antarctica and Greenland have been around for about one million years. Then, about 20,000 years ago, huge glaciers moved over North America, Europe, and parts of Asia. These glaciers were from a time known as the last Ice Age. As the glaciers moved, they eroded the land underneath them and changed it.

About 6,000 years ago, the Ice Age glaciers started melting. They left giant rocks and other debris in their place. You can still see signs of this Ice Age debris and erosion in many places.

GLACIERS

Glacial (GLAY-shul) erosion smoothed these large rocks. ▶

Glaciers carved canyons in the Rocky Mountains. ▶

Glaciers took along loose rock as they moved. When the glaciers melted, they left piles of this loose rock behind. ▶

They Made a Difference

Jean Louis Agassiz (A-guh-see) studied glaciers near his home in Switzerland. He saw that the glaciers left scratches in rocks and piles of debris. Then he saw those marks in places that did not have glaciers. Agassiz thought great glaciers had once covered the land during an ice age. Other scientists studied his work and decided that he was right.

CHAPTER 3

Water and Wind Erosion

▲ These rocks off the coast of Australia used to be part of the rocky cliffs.

Waves and wind cause constant erosion on the beach. The waves move back and forth against the shore and move the sand around. Storms with large waves can drag large amounts of sand into the water, causing fast beach erosion. Houses that once stood far from the water may wind up in it!

Waves wear away rocks near the beach, too. Over time, the rocks form caves, arches, and other shapes. This kind of erosion is slow, and goes on day after day.

Rivers move all the time. The moving water carries soil and rocks with it. Fast-moving rivers carry large rocks. Slower rivers carry small rocks, bits of sand, and clay. Over time, rivers can carve deep canyons through rock.

Rain beats down on rocks and earth, wearing them away. As rain flows downhill, it carries away bits of sand, soil, and other debris. Floods can cause much erosion very quickly.

7. Solve This

The Cape Hatteras (HA-tuh-rus) lighthouse in North Carolina was built in 1870. It was 1,500 feet (457 meters) from the ocean's edge. By 1999, the lighthouse stood only 150 feet (45 meters) from the ocean.
a. How much beach eroded between 1870 and 1999?
b. How many years did that erosion take?
c. Was the lighthouse farther away from the ocean after it was moved in 1999? Or was it farther away from the ocean in 1870? By how much?

Math ✓ Point

What information do you need to answer question c? Where can you find this information?

▲ This lighthouse was moved one-half mile from the ocean in 1999.

CHAPTER 3

Wind Erosion

Wind can change Earth in two ways. A strong wind can lift small bits of dust, soil, and sand high in the air. These bits may stay in the air for many miles. Then they drop to the ground far from where they started. Over time, wind can remove all of the earth from a place. Only rock is left.

Wind also pushes the soil and sand it carries against rocks. These bits of sand and soil rub against the rock. Some of the rock wears away.

▲ Wind erosion helped shape this rock, called a toadstool rock.

WATER AND WIND EROSION

Wind erodes land where there are no trees and plants to keep the earth in place. All over the world, farmers clear land. Wind can pick up the rich earth that the crops grow in and blow it away. Then only poor soil is left. To help stop erosion, farmers often plant trees or large bushes to act as windbreaks.

It's a Fact

In parts of Australia (au-STRALE-yuh), sheep, cattle, and rabbits have eaten most of the plants in open areas. The wind carries away the earth. A huge dust storm in Melbourne in 1983 moved about two million tons of dust and dirt.

▲ Windbreaks on these fields slow down the speed of the wind.

25

CHAPTER 3

Measuring Erosion

Scientists use special pins to measure earth erosion. They put the pins in the ground and leave the top of the pin sticking out. They mark the part of the pin that touches the ground. Scientists might check the pins every week or every month. If the mark on the pin is sticking up higher, they know the earth around the pin has eroded.

Everyday Science

Earthworms help keep soil from eroding. The worms dig holes in the ground. During a rain, the water flows into the ground. It does not erode the earth on top.

26

▲ eroded sandstone in Arizona

Rocks that are partly in the ground can be used to measure soil erosion. People paint a line around a rock, just above the spot where it meets the ground. As the soil erodes, it leaves an unpainted strip below the line. People measure the unpainted strip to find out how much the soil has eroded.

CHAPTER 3

Humans Cause Erosion, Too

People clear away trees and bushes to build houses, stores, and parking lots. Then there are no tree roots to hold the earth in place. Dry earth or mud can slide down a hill. Landslides and mudslides can cause huge damage and, sometimes, death.

▲ Haiti

People in the island country of Haiti (HAY-tee) cut down the forests to build their houses. The bare earth eroded. Landslides happen when it rains.

They Made a Difference

George Washington Carver studied plants and farming. Carver showed farmers how to plant crops that made the earth better. He told farmers to change their crops around, too. They stopped planting the same crop in the same place all the time. Carver's methods helped stop erosion.

✓Point Make Connections

Talk with people in your neighborhood who have gardens. Ask them how they prevent erosion.
What could you do to prevent or lessen the effects of erosion?

Conclusion

You have explored Earth's changes. Some are fast, while others are slow. You can only see some changes. Sometimes the hidden ones have the strongest effects.

You have discovered how scientists study and measure these changes. Their work helps us understand why and how Earth's surface changes. It can help us think of better ways to protect Earth and its people.

Perhaps someday you will be one of the scientists who track and measure Earth's changes!

Solve This — Answers

1. Page 8
17 minutes. Subtract 18 from 35.
Math Checkpoint
You need to know the time when the earthquake hit Australia and the time when it hit Massachusetts.

2. Page 9
100 times
Math Checkpoint
"Each number on the scale represents an earthquake that is 10 times stronger than the number just below it."

3. Page 14
about 180 feet
60 meters x 3 feet = 180 feet.
Math Checkpoint
It is an estimate. The problem says one meter is equal to about 3 feet. The question asks about how thick the glacier is in feet.

4. Page 15
a. 196 feet. 98 feet x 2 years = 196 feet.
b. 49 feet. Divide 98 feet by two, because six months is half a year. 98 feet divided by 2 = 49 feet.
Math Checkpoint
a. multiplication
b. division

5. Page 17
600 inches. 12 inches a year x 50 years = 600 inches
1,500 centimeters. 30 centimeters a year x 50 years = 1,500 centimeters

Math Checkpoint
It is an estimate. The problem says the glacier melts at about the same rate each year.

6. Page 18
a. 328 feet per day. 26,896 feet divided by 82 days = 328 feet per day.
b. 100 meters per day. 8,200 meters divided by 82 days = 100 meters per day.
Math Checkpoint
a. multiply 328 feet per day x 82 days = 26,896 feet
b. multiply 100 meters per day x 82 days = 8,200 meters

7. Page 23
a. 1,350 feet. 1,500 feet − 150 feet = 1,350 feet.
412 meters. 457 meters − 45 meters = 412 meters.
b. 129 years. 1999 − 1870 = 129 years.
c. The lighthouse was 1,140 feet farther from the ocean after it was moved in 1999. In 1870 it was 1,500 feet from the ocean. In 1999 it was moved one-half mile (2,640 feet) from the ocean. 2,640 feet − 1,500 feet = 1,140 feet.
Math Checkpoint
The photo caption says the lighthouse was moved one-half mile away from the ocean in 1999.

Glossary

debris	(deh-BREE) bits of rock and soil that are loosened by glaciers (page 19)
earthquake	(ERTH-kwake) a shaking or trembling of Earth's crust (page 2)
epicenter	(EH-pih-sen-ter) the place on the surface of Earth just above where the earthquake starts (page 6)
erosion	(ih-ROH-zhun) process that moves rocks, dirt, and sand from one place to another (page 2)
fault	(FAULT) the boundary between two tectonic plates; a crack in the earth caused by movement of the plates (page 5)
force	(FORS) a pushing or pulling action (page 2)
glacier	(GLAY-sher) a huge mass of ice and snow that moves under its own weight (page 2)
magnitude	(MAG-nih-tood) the strength or power of an earthquake (page 9)
plates	(PLATES) sections of Earth's crust (page 5)
seismograph	(SIZE-muh-graf) a machine that measures waves from an earthquake (page 7)
seismologist	(size-MAH-luh-jist) a scientist who studies earthquakes (page 7)
tsunami	(soo-NAH-mee) a huge wave often created by an earthquake (page 12)

Index

Agassiz, Jean Louis, 21
Antarctica, 2, 13, 20
Avouac, Jean-Philippe, 11
debris, 19–21
earthquake, 2–12
epicenter, 6, 12
erosion, 2, 19, 20–28
fault, 5, 7, 10
force, 2–3
galloping glacier, 18
glacier, 2–3, 13–21
ice sheet, 13
laser beams, 11, 16–17
magnitude, 9–12
plate, 5, 7, 10
Richter, Charles, 9
Richter scale, 9
satellite, 17
seismograph, 7–8, 11
seismologist, 7
tsunami, 12